Houghton
Mifflin
Harcourt

© Houghton Mifflin Harcourt Publishing Company • Cover Image Credits: (Goosander) ©Erich Kuchling/ Westend61/Corbis; (Covered bridge, New Hampshire) ©eye35/Alamy Images

Made in the United States
Text printed on 100%
recycled paper

Houghton
Mifflin
Harcourt

Printed in the U.S.A.

ISBN 978-0-544-34205-7

12 0928 22 21 20 19 18 17

4500661256   D E F G

Dear Students and Families,

Welcome to **Go Math!**, Grade 2! In this exciting mathematics program, there are hands-on activities to do and real-world problems to solve. Best of all, you will write your ideas and answers right in your book. In **Go Math!**, writing and drawing on the pages helps you think deeply about what you are learning, and you will really understand math!

By the way, all of the pages in your **Go Math!** book are made using recycled paper. We wanted you to know that you can Go Green with **Go Math!**

Sincerely,

The Authors

Made in the United States
Text printed on 100% recycled paper

GO MATH!

Authors

Juli K. Dixon, Ph.D.
Professor, Mathematics Education
University of Central Florida
Orlando, Florida

Edward B. Burger, Ph.D.
President, Southwestern University
Georgetown, Texas

Steven J. Leinwand
Principal Research Analyst
American Institutes for
    Research (AIR)
Washington, D.C.

Contributor

Rena Petrello
Professor, Mathematics
Moorpark College
Moorpark, California

Matthew R. Larson, Ph.D.
K-12 Curriculum Specialist for
    Mathematics
Lincoln Public Schools
Lincoln, Nebraska

Martha E. Sandoval-Martinez
Math Instructor
El Camino College
Torrance, California

English Language Learners Consultant

Elizabeth Jiménez
CEO, GEMAS Consulting
Professional Expert on English
    Learner Education
Bilingual Education and
    Dual Language
Pomona, California

Measurement and Data

 Critical Area Using standard units of measure

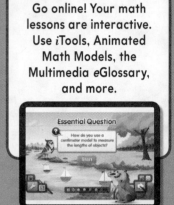

GO DIGITAL

Go online! Your math lessons are interactive. Use *i*Tools, Animated Math Models, the Multimedia *e*Glossary, and more.

Essential Question
How do you use a centimeter model to measure the lengths of objects?
Start

Chapter 9 Overview

In this chapter, you will explore and discover answers to the following **Essential Question**s:

• What are some of the methods and tools that can be used to estimate and measure length in metric units?

• What tools can be used to measure length in metric units and how do you use them?

• What metric units can be used to measure length and how do they compare with each other?

• If you know the length of one object, how can you estimate the length of another object?

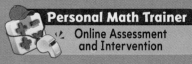

Personal Math Trainer
Online Assessment and Intervention

**FOR MORE PRACTICE
GO TO THE
Personal Math Trainer**

**Practice and
Homework**

Lesson Check and
Spiral Review in
every lesson

Length in Metric Units

A wind farm is a group of wind turbines used to make electricity. One way to measure the distance between two wind turbines is by counting footsteps. What is another way?

Name _____

Compare Lengths

1. Order the strings from shortest to longest.
   Write 1, 2, 3. (1.MD.A.1)

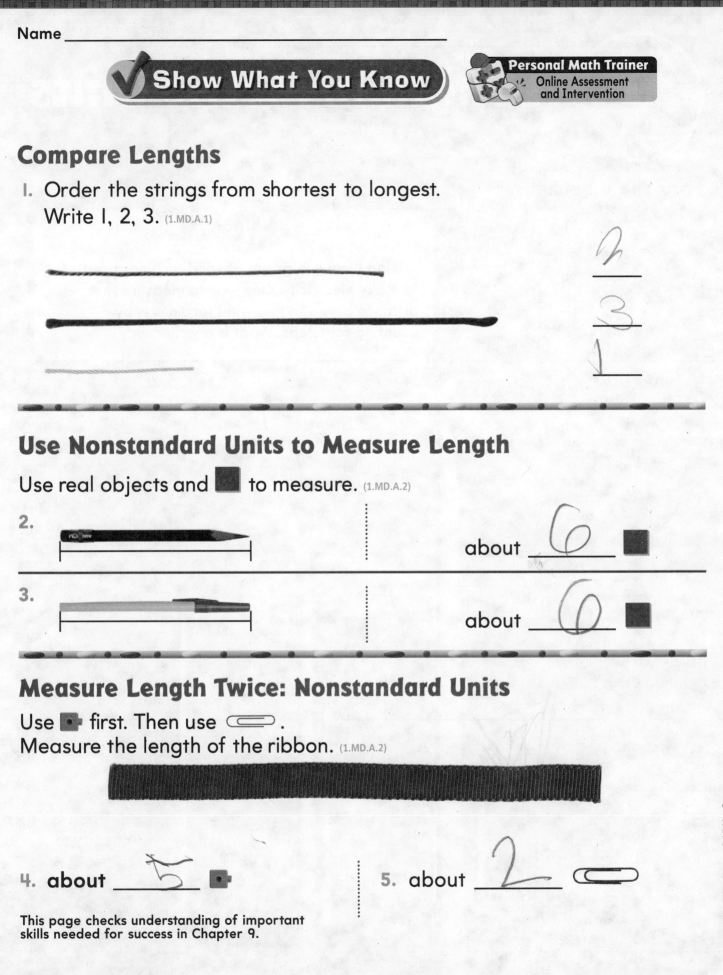

2

3

1

Use Nonstandard Units to Measure Length

Use real objects and ■ to measure. (1.MD.A.2)

2.   about ___6___ ■

3.   about ___6___ ■

Measure Length Twice: Nonstandard Units

Use ▪ first. Then use ⬭.
Measure the length of the ribbon. (1.MD.A.2)

4. about ___5___ ▪   5. about ___2___ ⬭

This page checks understanding of important
skills needed for success in Chapter 9.

Vocabulary Builder

Visualize It

Fill in the graphic organizer. Think of an object and write about how you can **measure** the **length** of that object.

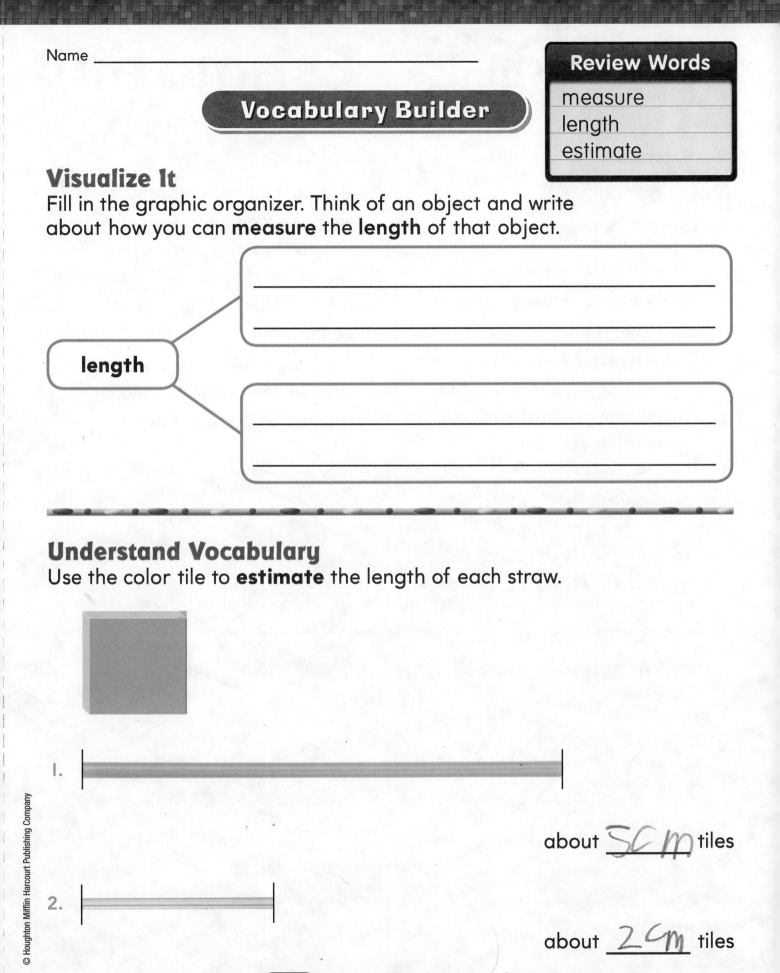

length

Understand Vocabulary

Use the color tile to **estimate** the length of each straw.

1.

about 5cm tiles

2.

about 2cm tiles

Chapter 4

Game

Estimating Length

Materials

- 12
- 12
- 15
- 15

Play with a partner.

1. Take turns choosing a picture. Find the real object.

2. Each player estimates the length of the object in cubes and then makes a cube train for his or her estimate.

3. Compare the cube trains to the length of the object. The player with the closer estimate puts a counter on the picture. If there is a tie, both players put a counter on the picture.

4. Repeat until all pictures are covered. The player with more counters on the board wins.

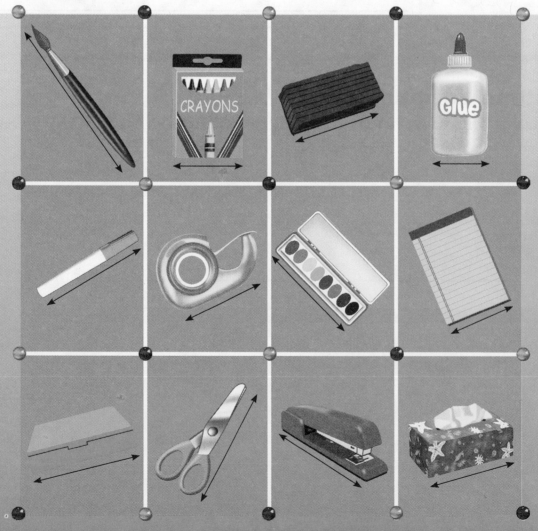

Chapter 9 Vocabulary

addend sumando 1	**centimeter** centímetro 6
compare comparar 8	**difference** diferencia 14
digit dígito 15	**estimate** estimación 21
meter (m) metro (m) 39	**sum** suma o total 59

This is 1 **centimeter**.

5 + 3 = 8

addends

5 − 3 = 2

difference

Compare the lengths of the pencil and the crayon.

The pencil is longer than the crayon. The crayon is shorter than the pencil.

An **estimate** is an amount that tells about how many.

0, 1, 2, 3, 4, 5, 6, 7, 8, and 9 are **digits**.

4 + 2 = 6

sum

1 **meter** is the same length as 100 centimeters.

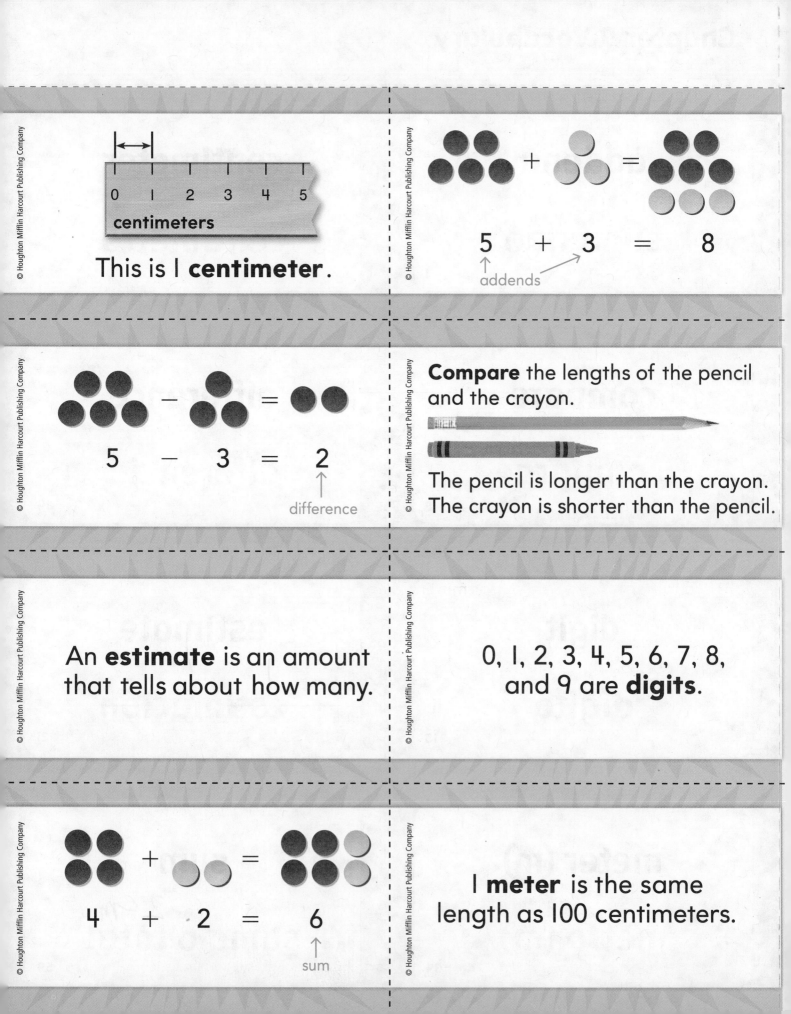

Make a Match

Word Box

addend

centimeter

compare

difference

digit

estimate

meter

sum

For 3 players

Materials

• 4 sets of word cards

How to Play

1. Every player is dealt 5 cards. Put the rest face-down in a draw pile.

2. Ask another player for a word card to match a word card you have.

   • If the player has the word card, he or she gives it to you. Put both cards in front of you. Take another turn.

   • If the player does not have the word card, take a card from the pile. If the word you get matches one you are holding, put both cards in front of you. Take another turn. If it does not match, your turn is over.

3. The game is over when one player has no cards left. The player with the most pairs wins.

The Write Way

Reflect

Choose one idea. Write about it in the space below.

- Compare a centimeter to a meter. Explain how they are alike and how they are different.

- Explain how you would find the length of this crayon in centimeters.

- How would you compare the length of a door to the length of a window in meters? Draw pictures and write to explain. Use another piece of paper for your drawing.

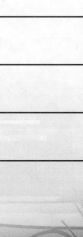

Name _____

Measure with a Centimeter Model

Essential Question How do you use a centimeter model to measure the lengths of objects?

Common Core Measurement and Data—
2.MD.A.1
MATHEMATICAL PRACTICES
MP5, MP6, MP8

Listen and Draw

Use ▪ to measure the length.

_____ 4 unit cubes

_____ unit cubes

_____ unit cubes

Math Talk MATHEMATICAL PRACTICES **5**

Use Tools Describe how to use unit cubes to measure an object's length.

HOME CONNECTION • Your child used unit cubes as an introduction to measurement of length before using metric measurement tools.

A unit cube is about 1 **centimeter** long.

About how many centimeters long is this string?

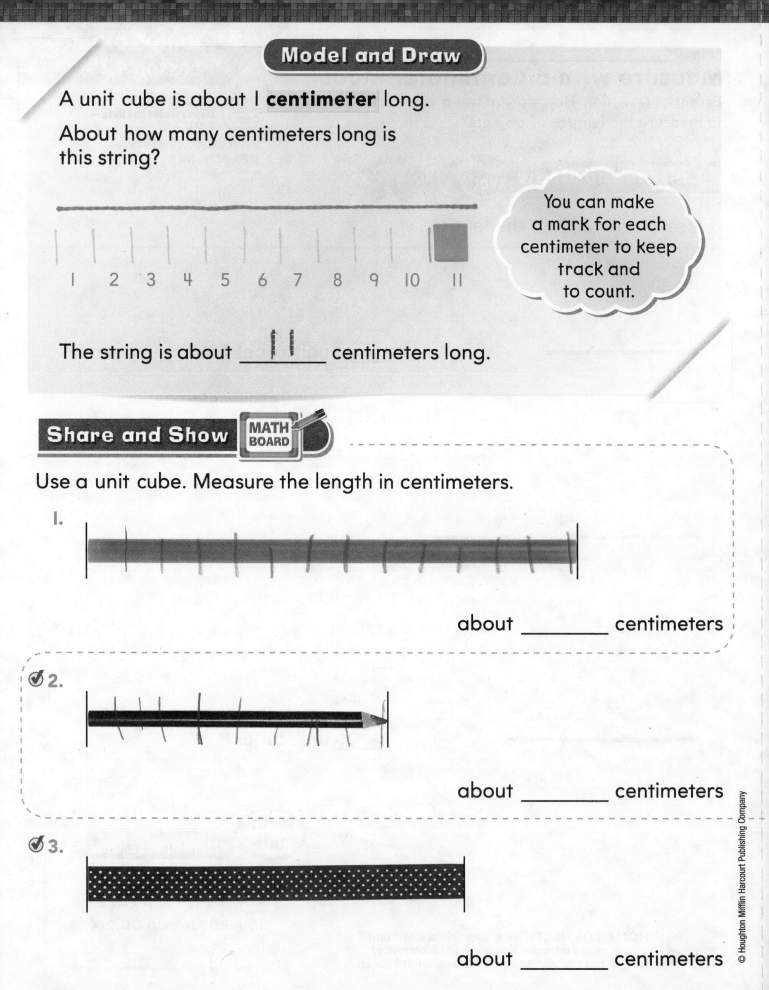

You can make a mark for each centimeter to keep track and to count.

The string is about _____ centimeters long.

Share and Show MATH BOARD

Use a unit cube. Measure the length in centimeters.

1.

about _____ centimeters

2.

about _____ centimeters

3.

about _____ centimeters

© Houghton Mifflin Harcourt Publishing Company

Name _____

Use a unit cube. Measure the length in centimeters.

4.

about __3½cn__ centimeters

5.

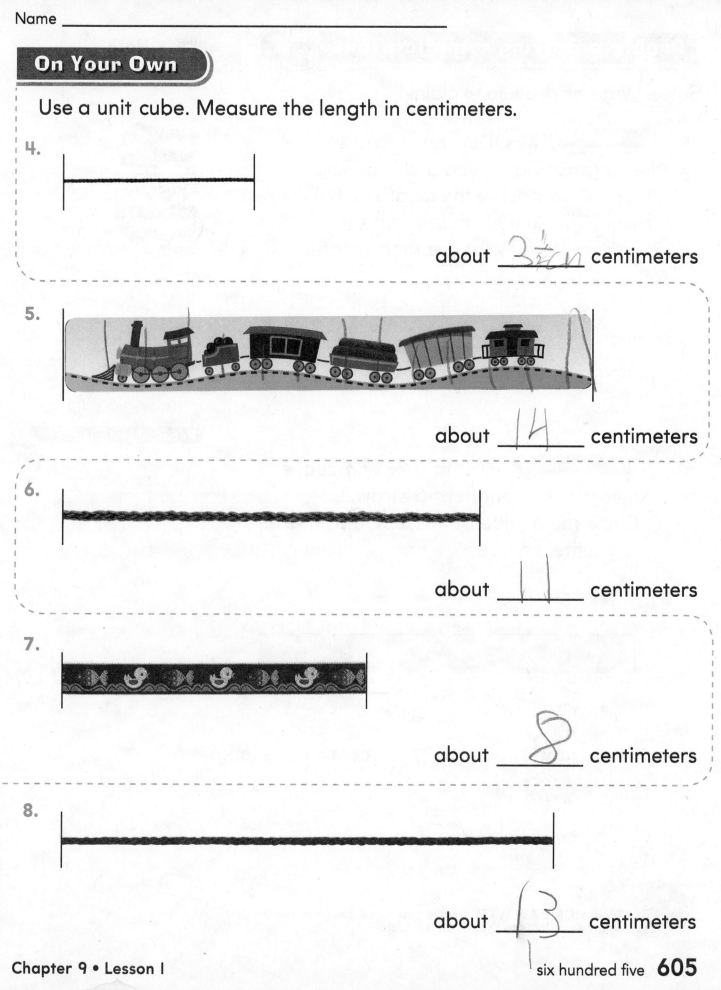

about __14__ centimeters

6.

about __11__ centimeters

7.

about __8__ centimeters

8.

about __13__ centimeters

Problem Solving • Applications (Real World)   WRITE Math

Solve. Write or draw to explain.

9. **THINK SMARTER** Mrs. Duncan measured the lengths of a crayon and a pencil. The pencil is double the length of the crayon. The sum of their lengths is 24 centimeters. What are their lengths?

crayon: _____

pencil: _____

Personal Math Trainer

10. **THINK SMARTER +** Marita uses unit cubes to measure the length of a straw.
Circle the number in the box that makes the sentence true.

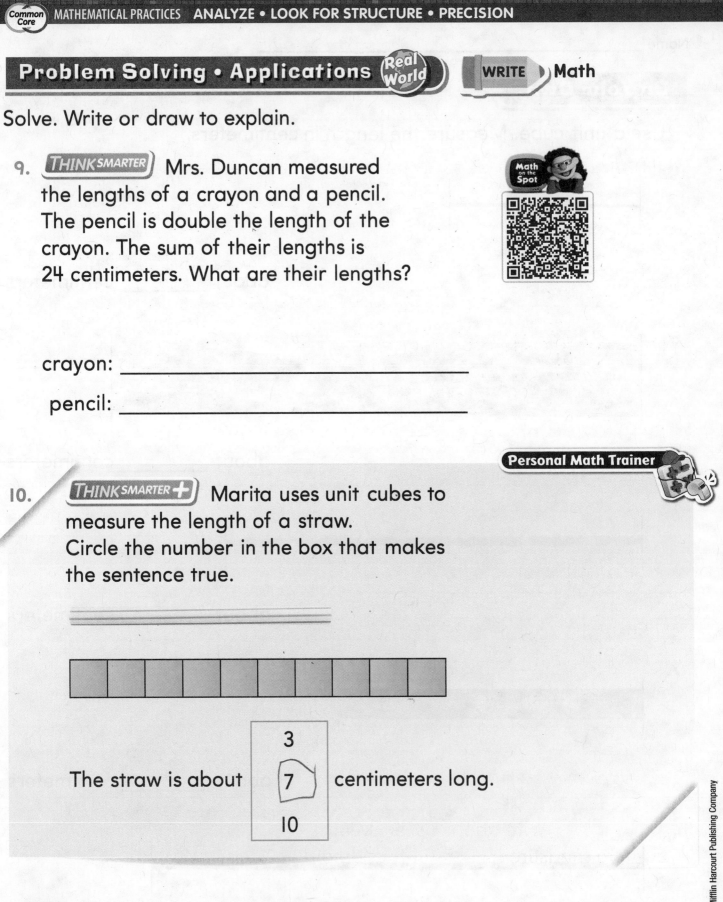

The straw is about
| 3 |
| 7 |
| 10 |
centimeters long.

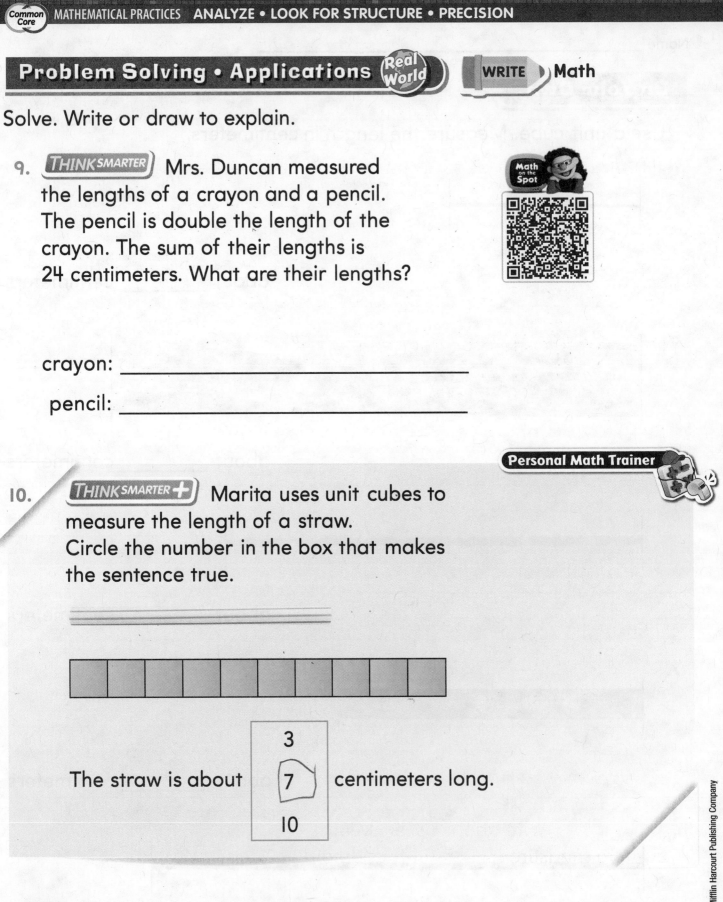

🏠 **TAKE HOME ACTIVITY** • Have your child compare the lengths of other objects to those in this lesson.

Name _____

Measure with a Centimeter Model

COMMON CORE STANDARD—2.MD.A.1
Measure and estimate lengths in standard units.

Use a unit cube. Measure the length in centimeters.

1.

about __6__ centimeters

2.

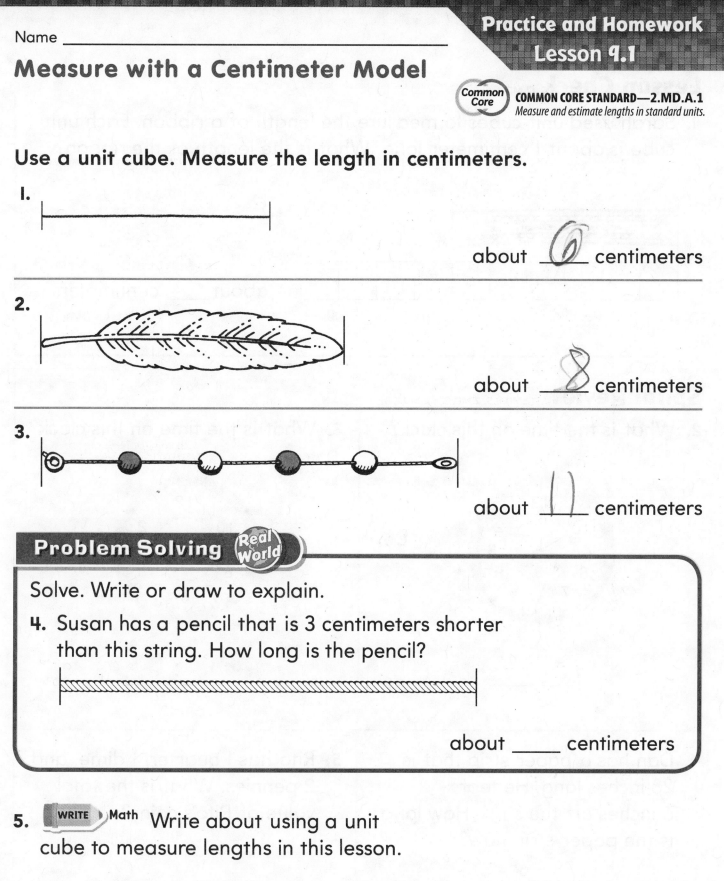

about __8__ centimeters

3.

about __11__ centimeters

Problem Solving · Real World

Solve. Write or draw to explain.

4. Susan has a pencil that is 3 centimeters shorter than this string. How long is the pencil?

about _____ centimeters

5. **WRITE** Math Write about using a unit cube to measure lengths in this lesson.

_____

_____

Lesson Check (2.MD.A.1)

1. Sarah used unit cubes to measure the length of a ribbon. Each unit cube is about 1 centimeter long. What is the length of the ribbon?

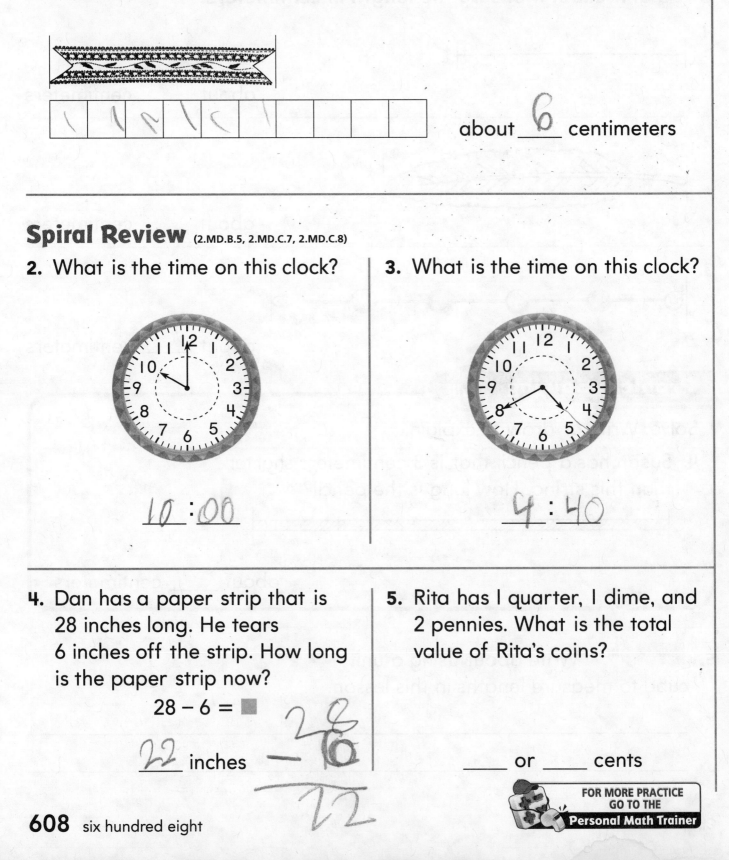

about __6__ centimeters

Spiral Review (2.MD.B.5, 2.MD.C.7, 2.MD.C.8)

2. What is the time on this clock?

10 : 00

3. What is the time on this clock?

4 : 40

4. Dan has a paper strip that is 28 inches long. He tears 6 inches off the strip. How long is the paper strip now?

$28 - 6 = \blacksquare$

__22__ inches

5. Rita has 1 quarter, 1 dime, and 2 pennies. What is the total value of Rita's coins?

____ or ____ cents

FOR MORE PRACTICE GO TO THE Personal Math Trainer

Name _____

Estimate Lengths in Centimeters

Essential Question How do you use known lengths to estimate unknown lengths?

Common Core   Measurement and Data—2.MD.A.3
MATHEMATICAL PRACTICES
MP1, MP6, MP7

Listen and Draw Real World   Hands On

Find three classroom objects that are shorter than your 10-centimeter strip. Draw the objects. Write estimates for their lengths.

phone stick

about _25_ centimeters

paper

about _44_ centimeters

about _____ centimeters

Math Talk
MATHEMATICAL PRACTICES 6

Which object has a length closest to 10 centimeters? Explain.

HOME CONNECTION • Your child used a 10-centimeter strip of paper to practice estimating the lengths of some classroom objects.

This pencil is about 10 centimeters long.
Which is the most reasonable estimate
for the length of the ribbon?

7 centimeters

13 centimeters

20 centimeters

The ribbon is longer
than the pencil.
7 centimeters is not
reasonable.

The ribbon is not
twice as long as the pencil.
20 centimeters is not
reasonable.

The ribbon is a little longer than the pencil.
So, 13 centimeters is the most reasonable estimate.

Share and Show   MATH BOARD

1. The yarn is about 5 centimeters long. Circle the
best estimate for the length of the crayon.

10 centimeters

15 centimeters

20 centimeters

2. The string is about 12 centimeters long.
Circle the best estimate for the length of the straw.

3 centimeters

7 centimeters

11 centimeters

Name _____

On Your Own

3. The rope is about 8 centimeters long. Circle the
best estimate for the length of the paper clip.

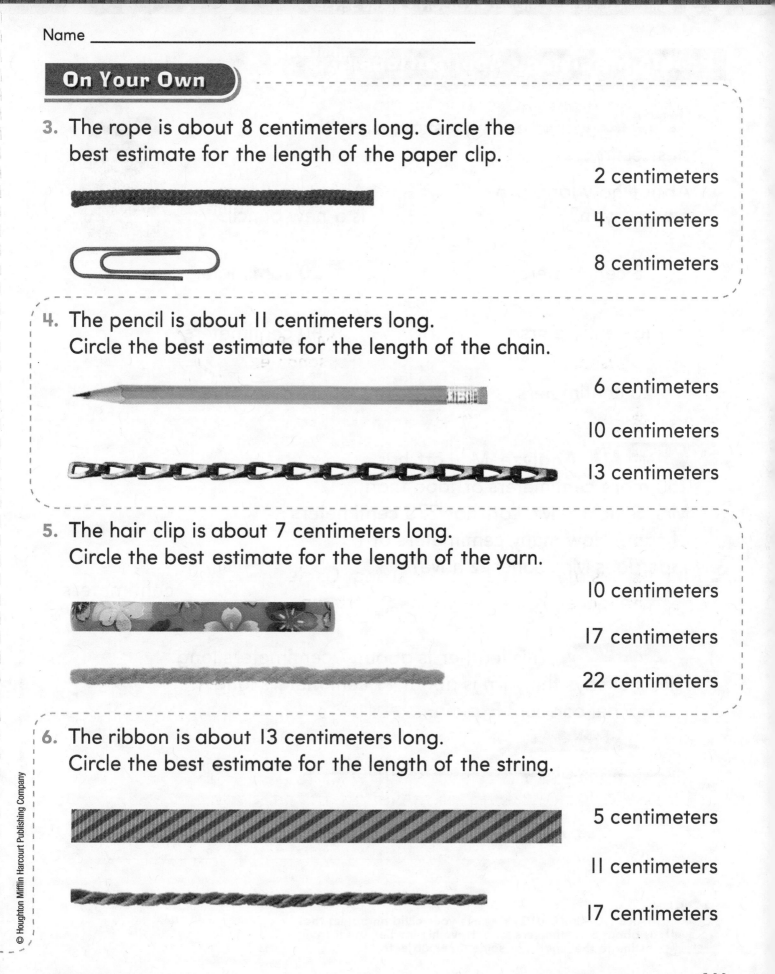

2 centimeters

4 centimeters

8 centimeters

4. The pencil is about 11 centimeters long.
Circle the best estimate for the length of the chain.

6 centimeters

10 centimeters

13 centimeters

5. The hair clip is about 7 centimeters long.
Circle the best estimate for the length of the yarn.

10 centimeters

17 centimeters

22 centimeters

6. The ribbon is about 13 centimeters long.
Circle the best estimate for the length of the string.

5 centimeters

11 centimeters

17 centimeters

© Houghton Mifflin Harcourt Publishing Company

Chapter 9 • Lesson 2

six hundred eleven **611**

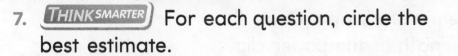

Problem Solving • Applications (Real World)   WRITE Math

7. **THINK SMARTER** For each question, circle the best estimate.

About how long is a
new crayon?

    5 centimeters

    10 centimeters

    20 centimeters

About how long
is a new pencil?

    20 centimeters

    40 centimeters

    50 centimeters

8. **MATHEMATICAL PRACTICE ①** **Analyze** Mr. Lott has 250 more centimeters of tape than Mrs. Sanchez. Mr. Lott has 775 centimeters of tape. How many centimeters of tape does Mrs. Sanchez have?

_____ centimeters

9. **THINK SMARTER** This feather is about 7 centimeters long. Rachel says the yarn is about 14 centimeters long. Is Rachel correct? Explain.

_____

_____

TAKE HOME ACTIVITY • Give your child an object that is about 5 centimeters long. Have him or her use it to estimate the lengths of some other objects.

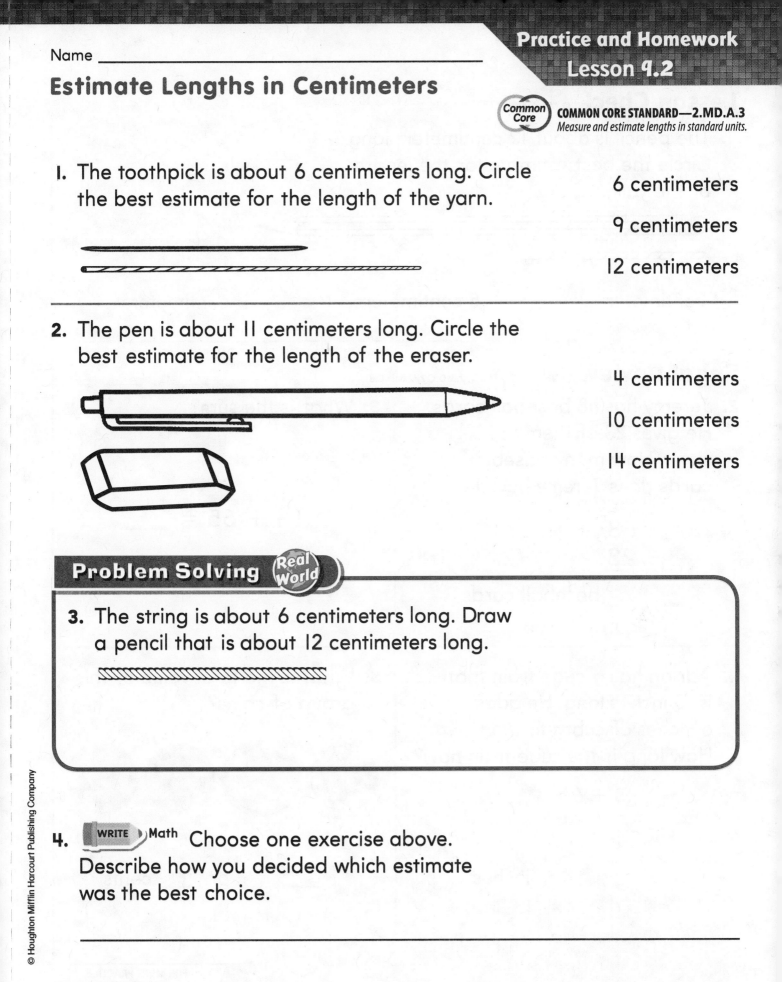

Name _____

Estimate Lengths in Centimeters

Common Core   **COMMON CORE STANDARD—2.MD.A.3**
Measure and estimate lengths in standard units.

1. The toothpick is about 6 centimeters long. Circle the best estimate for the length of the yarn.

   6 centimeters

   9 centimeters

   12 centimeters

2. The pen is about 11 centimeters long. Circle the best estimate for the length of the eraser.

   4 centimeters

   10 centimeters

   14 centimeters

Problem Solving Real World

3. The string is about 6 centimeters long. Draw a pencil that is about 12 centimeters long.

4. **WRITE** Math Choose one exercise above. Describe how you decided which estimate was the best choice.

Lesson Check (2.MD.A.3)

1. The pencil is about 12 centimeters long.
Circle the best estimate for the length
of the yarn.

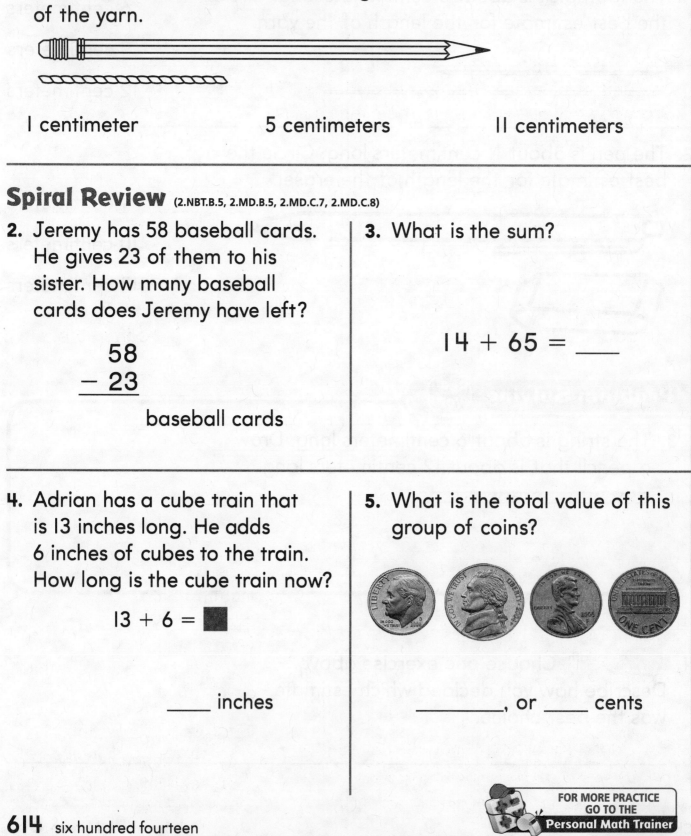

I centimeter   5 centimeters   II centimeters

Spiral Review (2.NBT.B.5, 2.MD.B.5, 2.MD.C.7, 2.MD.C.8)

2. Jeremy has 58 baseball cards.
He gives 23 of them to his
sister. How many baseball
cards does Jeremy have left?

$$\begin{array}{r} 58 \\ -\ 23 \\ \hline \end{array}$$

_____ baseball cards

3. What is the sum?

$$14 + 65 = \underline{\quad}$$

4. Adrian has a cube train that
is 13 inches long. He adds
6 inches of cubes to the train.
How long is the cube train now?

$$13 + 6 = \blacksquare$$

_____ inches

5. What is the total value of this
group of coins?

_____, or _____ cents

FOR MORE PRACTICE
GO TO THE
Personal Math Trainer

Name _____

Measure with a Centimeter Ruler

Essential Question How do you use a centimeter ruler to measure lengths?

Common Core **Measurement and Data—**
2.MD.A.1
MATHEMATICAL PRACTICES
MP3, MP5, MP6

Listen and Draw Real World

Find three small objects in the classroom.
Use unit cubes to measure their lengths.
Draw the objects and write their lengths.

about _____ centimeters

about _____ centimeters

about _____ centimeters

Math Talk MATHEMATICAL PRACTICES 3

Apply
Describe how the three lengths compare. Which object is shortest?

HOME CONNECTION • Your child used unit cubes to measure the lengths of some classroom objects as an introduction to measuring lengths in centimeters.

Model and Draw

What is the length of the crayon to the nearest centimeter?

Remember: Line up the left edge of the object with the zero mark on the ruler.

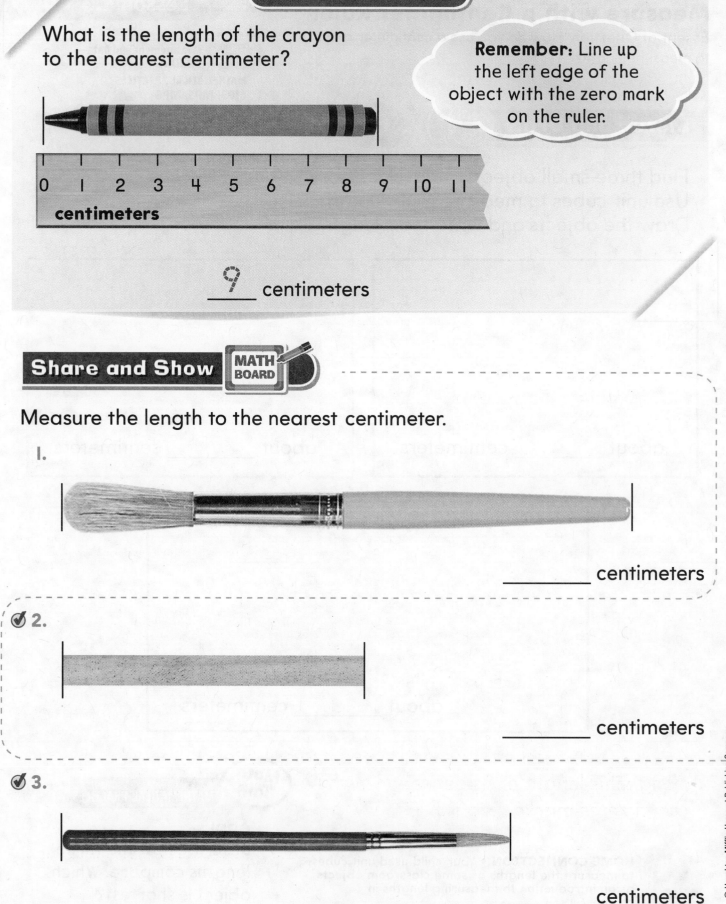

_9_ centimeters

Share and Show MATH BOARD

Measure the length to the nearest centimeter.

1.

_____ centimeters

☑ 2.

_____ centimeters

☑ 3.

_____ centimeters

Name _____

Measure the length to the nearest centimeter.

4.

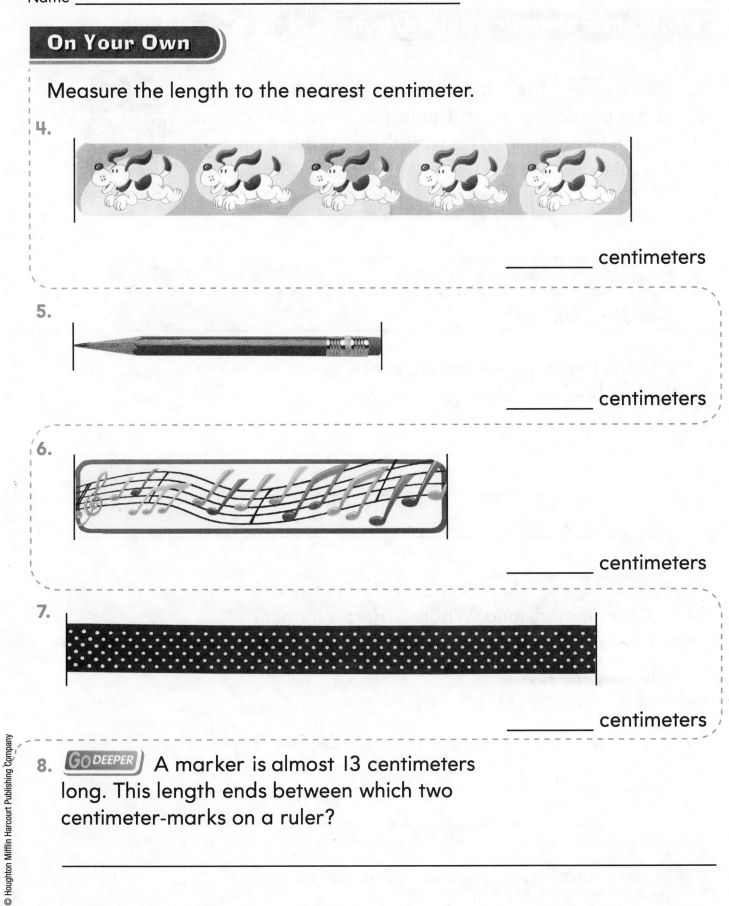

_____ centimeters

5.

_____ centimeters

6.

_____ centimeters

7.

_____ centimeters

8. **GO DEEPER** A marker is almost 13 centimeters long. This length ends between which two centimeter-marks on a ruler?

_____

_____

Problem Solving • Applications Real World   WRITE Math

9. **THINK SMARTER** The crayon was on the table next to the centimeter ruler. The left edge of the crayon was not lined up with the zero mark on the ruler.

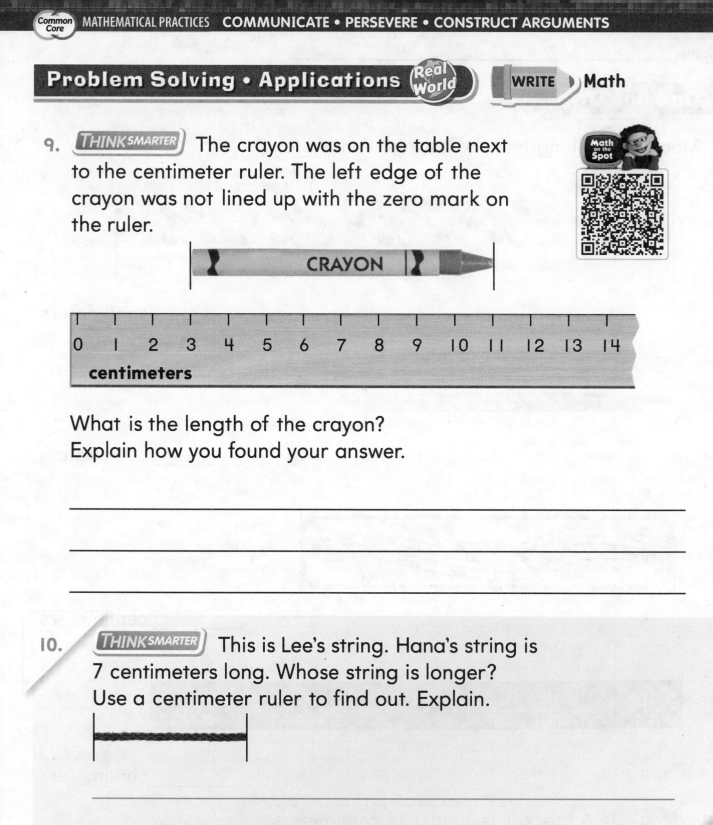

What is the length of the crayon?
Explain how you found your answer.

_____

_____

_____

10. **THINK SMARTER** This is Lee's string. Hana's string is 7 centimeters long. Whose string is longer? Use a centimeter ruler to find out. Explain.

_____

_____

TAKE HOME ACTIVITY • Have your child measure the lengths of some objects using a centimeter ruler.

Measure with a Centimeter Ruler

Common Core COMMON CORE STANDARD—2.MD.A.1
Measure and estimate lengths in standard units.

Measure the length to the nearest centimeter.

1.

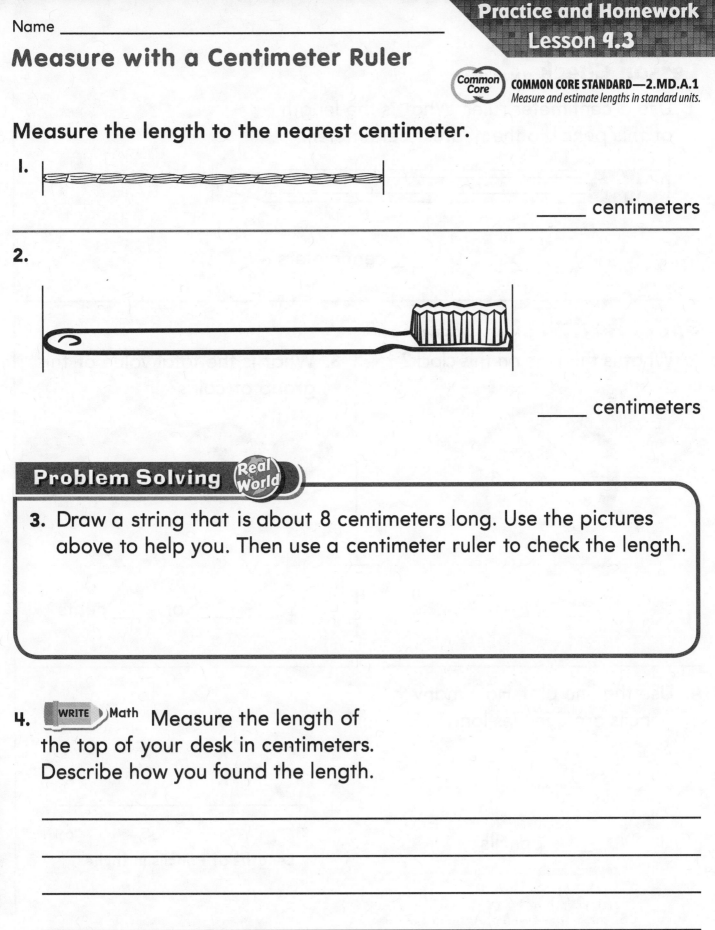

_____ centimeters

2.

_____ centimeters

Problem Solving Real World

3. Draw a string that is about 8 centimeters long. Use the pictures above to help you. Then use a centimeter ruler to check the length.

4. **WRITE** Math Measure the length of the top of your desk in centimeters. Describe how you found the length.

_____

_____

_____

_____

Lesson Check (2.MD.A.1)

1. Use a centimeter ruler. What is the length of this pencil to the nearest centimeter?

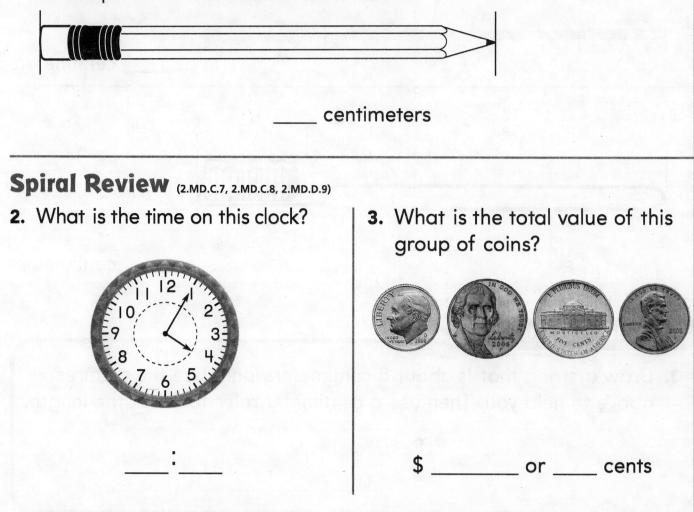

_____ centimeters

Spiral Review (2.MD.C.7, 2.MD.C.8, 2.MD.D.9)

2. What is the time on this clock?

_____ : _____

3. What is the total value of this group of coins?

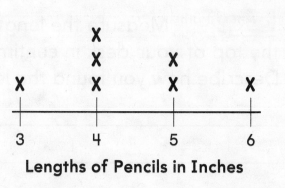

$ _____ or _____ cents

4. Use the line plot. How many pencils are 5 inches long?

_____ pencils

```
                        X
                        X         X
          X             X         X              X
          +-------------+---------+--------------+
          3             4         5              6
```
Lengths of Pencils in Inches

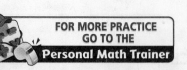

FOR MORE PRACTICE
GO TO THE
Personal Math Trainer

Name _____

Problem Solving • Add and Subtract Lengths

Essential Question How can drawing a diagram help when solving problems about lengths?

Common Core — Measurement and Data—
2.MD.B.6, 2.MD.B.5
MATHEMATICAL PRACTICES
MP1, MP2, MP4

Nate had 23 centimeters of string.
He gave 9 centimeters of string to Myra.
How much string does Nate have now?

Unlock the Problem (Real World)

What do I need to find?

how much string
Nate has now

What information do I need to use?

Nate had _____ centimeters of string.

He gave _____ centimeters of string to Myra.

Show how to solve the problem.

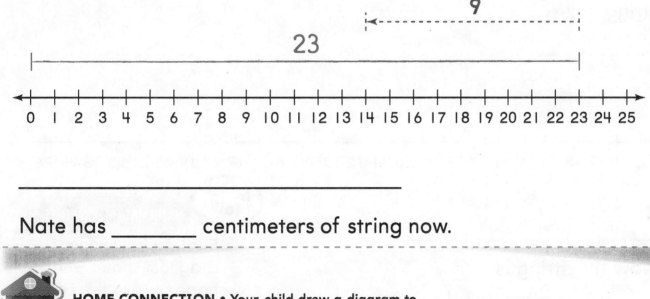

_____

Nate has _____ centimeters of string now.

HOME CONNECTION • Your child drew a diagram to represent a problem about lengths. The diagram can be used to choose the operation for solving the problem.

© Houghton Mifflin Harcourt Publishing Company

Try Another Problem

Draw a diagram. Write a number sentence using a ▇ for the unknown number. Then solve.

- What do I need to find?
- What information do I need to use?

1. Ellie has a ribbon that is 12 centimeters long. Gwen has a ribbon that is 9 centimeters long. How many centimeters of ribbon do they have?

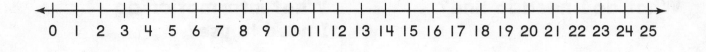

_____

They have _____ centimeters of ribbon.

2. A string is 24 centimeters long. Justin cuts 8 centimeters off. How long is the string now?

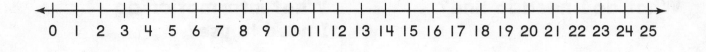

_____

Now the string is

_____ centimeters long.

Math Talk MATHEMATICAL PRACTICES 4

Explain how your diagram shows what happened in the first problem.

Name _____

Draw a diagram. Write a number sentence using a ▪ for the unknown number. Then solve.

☑**3.** A chain of paper clips is 18 centimeters long. Sondra adds 6 centimeters of paper clips to the chain. How long is the chain now?

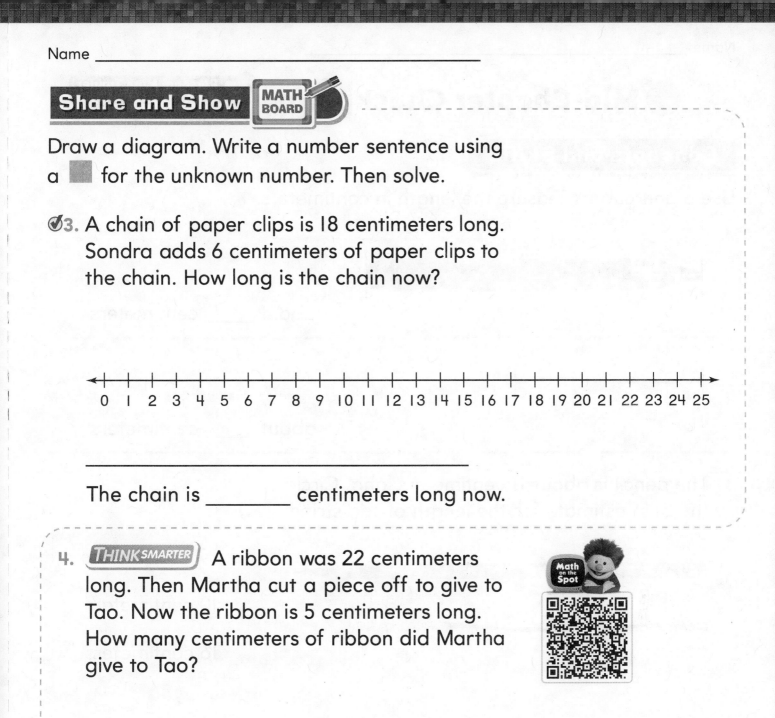

|←+—+→|
0 1 2 3 4 5 6 7 8 9 10 11 12 13 14 15 16 17 18 19 20 21 22 23 24 25

_____

The chain is _____ centimeters long now.

4. THINK SMARTER   A ribbon was 22 centimeters long. Then Martha cut a piece off to give to Tao. Now the ribbon is 5 centimeters long. How many centimeters of ribbon did Martha give to Tao?

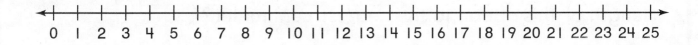

|←+—+→|
0 1 2 3 4 5 6 7 8 9 10 11 12 13 14 15 16 17 18 19 20 21 22 23 24 25

Martha gave _____ centimeters of ribbon to Tao.

TAKE HOME ACTIVITY • Have your child explain how he or she used a diagram to solve one problem in this lesson.

✓ Mid-Chapter Checkpoint

Personal Math Trainer
Online Assessment and Intervention

Concepts and Skills

Use a unit cube. Measure the length in centimeters. (2.MD.A.1)

1.

about _____ centimeters

2.

about _____ centimeters

3. The pencil is about 11 centimeters long. Circle
the best estimate for the length of the string. (2.MD.A.3)

7 centimeters

10 centimeters

16 centimeters

4. **THINK SMARTER** Use a centimeter ruler. What is the
length of this ribbon to the nearest centimeter? (2.MD.A.1)

_____ centimeters

Problem Solving • Add and Subtract Lengths

 Common Core — **COMMON CORE STANDARDS—2.MD.B.6, 2.MD.B.5** *Relate addition and subtraction to length.*

Draw a diagram. Write a number sentence using a ▢ for the unknown number. Then solve.

1. A straw is 20 centimeters long. Mr. Jones cuts 8 centimeters off the straw. How long is the straw now?

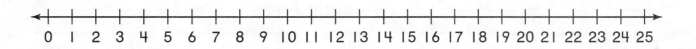

_____

The straw is _____ centimeters long now.

2. **WRITE** Math Draw and describe a diagram for a problem about the total length of two ribbons, 13 centimeters long and 5 centimeters long.

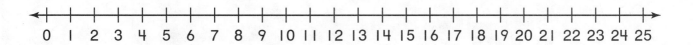

Lesson Check (2.MD.B.6, 2.MD.B.5)

1. Tina has a paper clip chain that is 25 centimeters long. She takes off 8 centimeters of the chain. How long is the chain now?

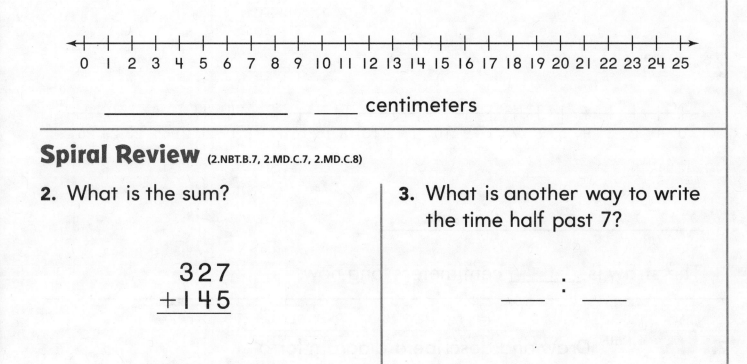

_____ _____ centimeters

Spiral Review (2.NBT.B.7, 2.MD.C.7, 2.MD.C.8)

2. What is the sum?

$$\begin{array}{r} 327 \\ +145 \\ \hline \end{array}$$

3. What is another way to write the time half past 7?

_____ : _____

4. Molly has these coins in her pocket. How much money does she have in her pocket?

_____ or _____ cents

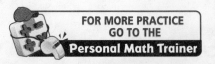

FOR MORE PRACTICE
GO TO THE
Personal Math Trainer

Name _____

Centimeters and Meters

Essential Question How is measuring in meters different from measuring in centimeters?

 Measurement and Data—
2.MD.A.2
MATHEMATICAL PRACTICES
MP1, MP5, MP7

Listen and Draw Real World · Hands On

Draw or write to describe how you did each measurement.

1st measurement

2nd measurement

Math Talk MATHEMATICAL PRACTICES 1

Describe how the lengths of the yarn and the sheet of paper are different.

FOR THE TEACHER • Have each small group use a 1-meter piece of yarn to measure a distance marked on the floor with masking tape. Then have them measure the same distance using a sheet of paper folded in half lengthwise.

Chapter 9

six hundred twenty-seven **627**

Model and Draw

I **meter** is the same as 100 centimeters.

The real door is about 200 centimeters tall.
The real door is also about 2 meters tall.

Share and Show MATH BOARD

Measure to the nearest centimeter.
Then measure to the nearest meter.

Find the real object.	Measure.
chair I.	_____ centimeters _____ meters
teacher's desk ☑ 2.	_____ centimeters _____ meters
wall ☑ 3.	_____ centimeters _____ meters

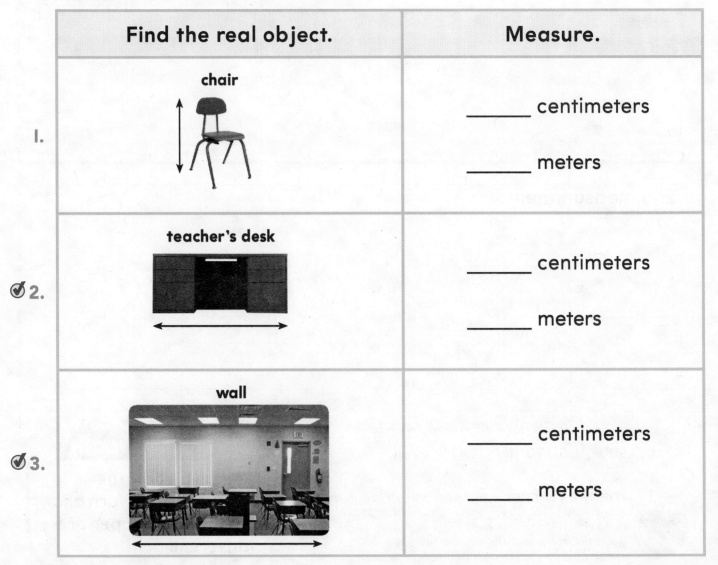

Name _____

On Your Own

Measure to the nearest centimeter.
Then measure to the nearest meter.

Find the real object.	Measure.
4. chalkboard 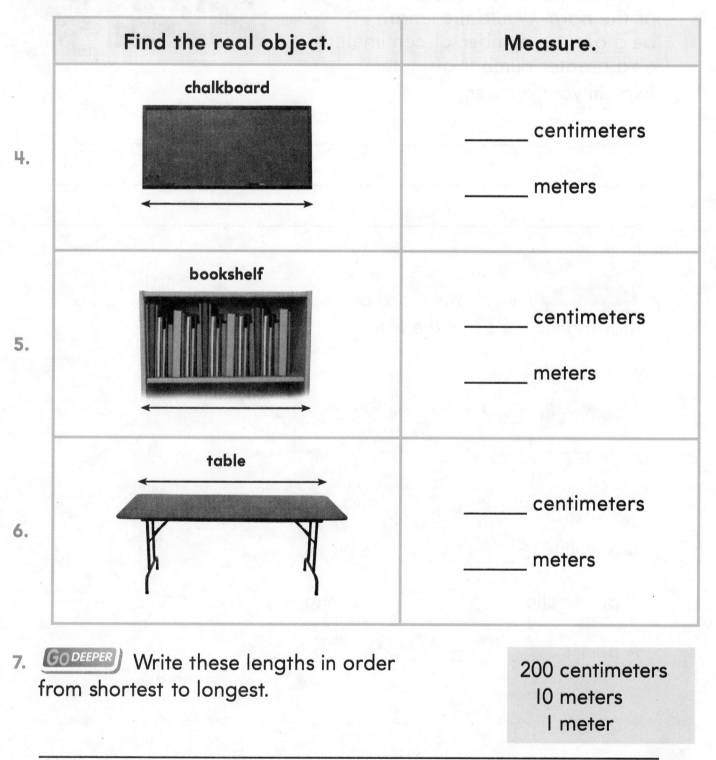	_____ centimeters _____ meters
5. bookshelf	_____ centimeters _____ meters
6. table	_____ centimeters _____ meters

7. GO DEEPER Write these lengths in order
from shortest to longest.

> 200 centimeters
> 10 meters
> 1 meter

_____

Problem Solving • Applications Real World WRITE Math

8. **THINK SMARTER** Mr. Ryan walked next to a barn. He wants to measure the length of the barn. Would the length be a greater number of centimeters or a greater number of meters? Explain your answer.

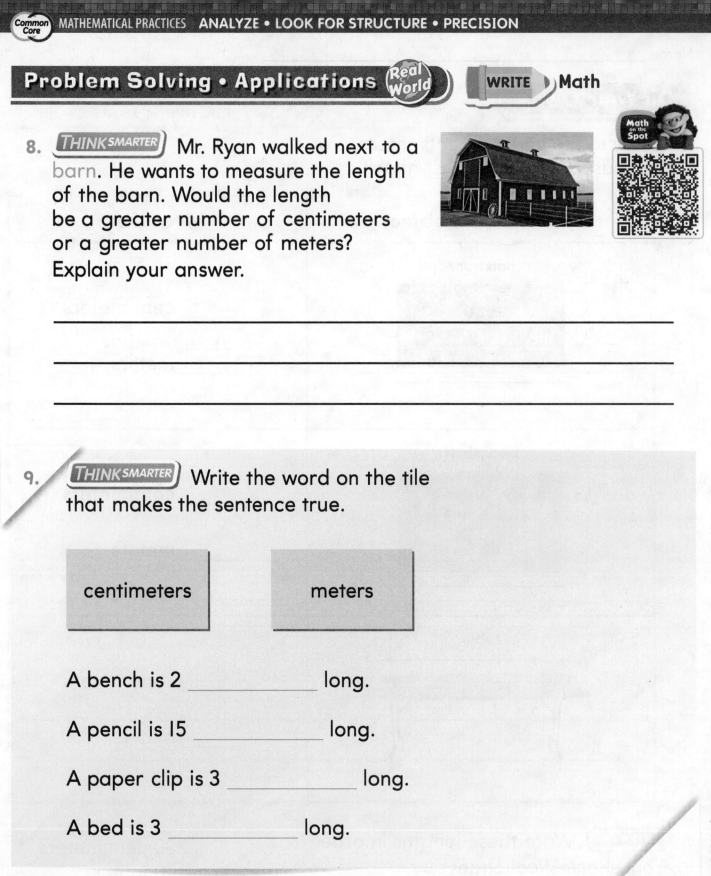

_____

_____

_____

9. **THINK SMARTER** Write the word on the tile that makes the sentence true.

centimeters	meters

A bench is 2 _____ long.

A pencil is 15 _____ long.

A paper clip is 3 _____ long.

A bed is 3 _____ long.

TAKE HOME ACTIVITY • Have your child describe how centimeters and meters are different.

Name _____

Centimeters and Meters

COMMON CORE STANDARD—2.MD.A.2
Measure and estimate lengths in standard units.

**Measure to the nearest centimeter.
Then measure to the nearest meter.**

Find the real object.	Measure.
1. bookcase 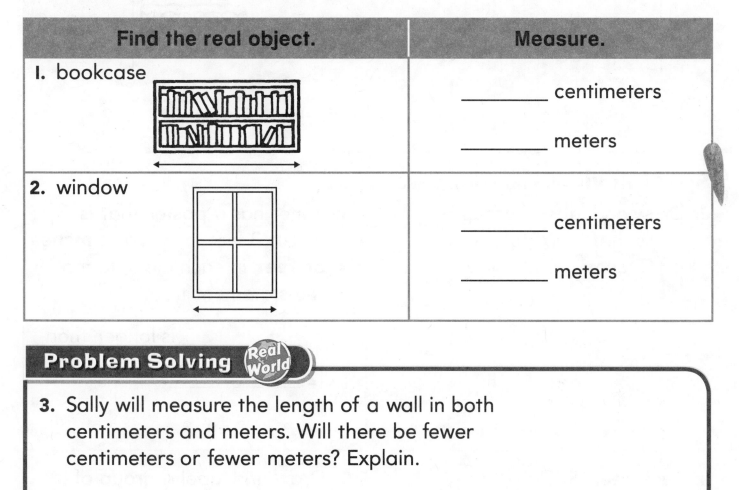	_____ centimeters _____ meters
2. window	_____ centimeters _____ meters

Problem Solving Real World

3. Sally will measure the length of a wall in both
centimeters and meters. Will there be fewer
centimeters or fewer meters? Explain.

_____

_____

_____

4. **WRITE** Math Would you measure the length of a bench
in centimeters or in meters? Explain your choice.

_____

Lesson Check (2.MD.A.2)

I. Use a centimeter ruler. What is the length of the toothbrush to the nearest centimeter?

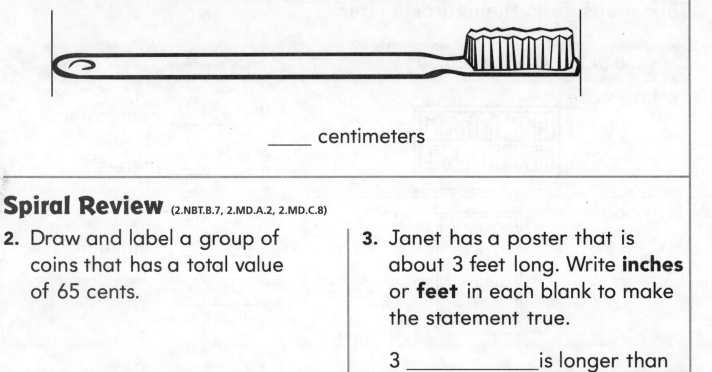

_____ centimeters

Spiral Review (2.NBT.B.7, 2.MD.A.2, 2.MD.C.8)

2. Draw and label a group of coins that has a total value of 65 cents.

3. Janet has a poster that is about 3 feet long. Write **inches** or **feet** in each blank to make the statement true.

3 _____ is longer than

12 _____ .

4. Last week, 483 children checked books out from the library. This week, only 162 children checked books out from the library. How many children checked out library books in the last two weeks?

483
+ 162
_____

5. Draw and label a group of coins that has a total value of $1.00.

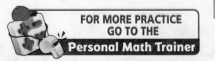

FOR MORE PRACTICE
GO TO THE
Personal Math Trainer

Name _____

Estimate Lengths in Meters

Essential Question How do you estimate the lengths of objects in meters?

Common Core · Measurement and Data—
2.MD.A.3
MATHEMATICAL PRACTICES
MP6, MP7

Listen and Draw (Real World)

Find an object that is about 10 centimeters long.
Draw and label it.

Is there a classroom object that is about
50 centimeters long? Draw and label it.

FOR THE TEACHER • Provide a collection of
objects for children to choose from. Above the
table of displayed objects, draw and label a
10-centimeter line segment and a 50-centimeter
line segment.

Math Talk — MATHEMATICAL PRACTICES 6

Describe how the
lengths of the two real
objects compare.

Estimate. About how many meter sticks will match the width of a door?

A 1-meter measuring stick is about 100 centimeters long.

about _____ meters

Share and Show MATH BOARD

Find the real object.
Estimate its length in meters.

☑ 1. bookshelf

about _____ meters

☑ 2. bulletin board

about _____ meters

On Your Own

Find the real object.
Estimate its length in meters.

3. teacher's desk

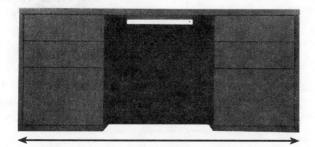

about _____ meters

4. wall

about _____ meters

5. window

about _____ meters

6. chalkboard

about _____ meters

Problem Solving • Applications

WRITE Math

7. **THINK SMARTER** In meters, estimate the distance from your teacher's desk to the door of your classroom.

about _____ meters

Explain how you made your estimate.

_____

_____

_____

_____

8. **THINK SMARTER** Estimate the length of an adult's bicycle. Fill in the bubble next to all the sentences that are true.

○ The bicycle is about 2 meters long.

○ The bicycle is about 200 centimeters long.

○ The bicycle is less than I meter long.

○ The bicycle is about 2 centimeters long.

○ The bicycle is more than 200 meters long.

TAKE HOME ACTIVITY • With your child, estimate the lengths of some objects in meters.

Estimate Lengths in Meters

Common Core
COMMON CORE STANDARD—2.MD.A.3
Measure and estimate lengths in standard units.

Find the real object.
Estimate its length in meters.

1. poster

about _____ meters

2. chalkboard

about _____ meters

Problem Solving · Real World

3. Barbara and Luke each placed 2 meter sticks
   end-to-end along the length of a large table.
   About how long is the table?

about _____ meters

4. WRITE · Math Choose one object from above.
   Describe how you estimated its length.

_____

_____

Lesson Check (2.MD.A.3)

1. What is the best estimate for the length of a real baseball bat?

_____ meter

2. What is the best estimate for the length of a real couch?

_____ meters

Spiral Review (2.MD.A.1, 2.MD.C.8)

3. Sara has two $1 bills, 3 quarters, and 1 dime. How much money does she have?

$ _____

4. Use an inch ruler. What is the length of this straw to the nearest inch?

_____ inches

5. Scott has this money in his pocket. What is the total value of this money?

$ _____

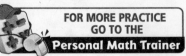

FOR MORE PRACTICE
GO TO THE
Personal Math Trainer

Name _____

Measure and Compare Lengths

Essential Question How do you find the difference
between the lengths of two objects?

Common Core **Measurement and Data—**
2.MD.A.4
MATHEMATICAL PRACTICES
MP1, MP2, MP6

Listen and Draw Real World   Hands On

Measure and record each length.

_____ centimeters

_____ centimeters

© Houghton Mifflin Harcourt Publishing Company

HOME CONNECTION • Your child measured
these lengths as an introduction to measuring
and then comparing lengths.

Math Talk MATHEMATICAL PRACTICES 6

Name a classroom
object that is longer
than the paintbrush.
Explain how you know.

Chapter 9

How much longer is the pencil than the crayon?

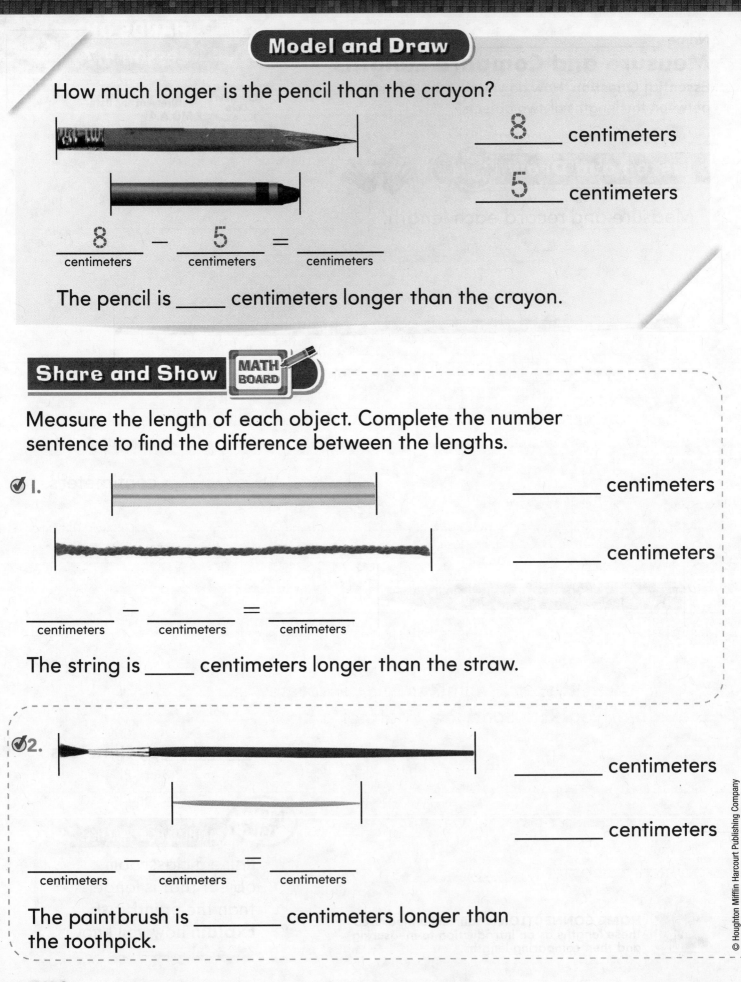

_____8_____ centimeters

_____5_____ centimeters

__8__ − __5__ = _____
centimeters   centimeters   centimeters

The pencil is _____ centimeters longer than the crayon.

Share and Show MATH BOARD

Measure the length of each object. Complete the number sentence to find the difference between the lengths.

☑ 1.

_____ centimeters

_____ centimeters

_____ − _____ = _____
centimeters   centimeters   centimeters

The string is _____ centimeters longer than the straw.

☑ 2.

_____ centimeters

_____ centimeters

_____ − _____ = _____
centimeters   centimeters   centimeters

The paintbrush is _____ centimeters longer than the toothpick.

Name _____

Measure the length of each object. Complete the number sentence to find the difference between the lengths.

3.

_____ centimeters

_____ centimeters

_____ − _____ = _____
centimeters   centimeters   centimeters

The yarn is _____ centimeters longer than the crayon.

4.

_____ centimeters

_____ centimeters

_____ − _____ = _____
centimeters   centimeters   centimeters

The string is _____ centimeters longer than the paper clip.

5. **THINK SMARTER** Use a centimeter ruler. Measure the length of your desk and the length of a book.

desk: _____ centimeters

book: _____ centimeters

Which is shorter? _____

How much shorter is it? _____

Problem Solving • Applications 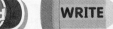 Math

6. Mark has a rope that is 23 centimeters long. He cuts 15 centimeters off. What is the length of the rope now?

_____ centimeters

7. The yellow ribbon is 15 centimeters longer than the green ribbon. The green ribbon is 29 centimeters long. What is the length of the yellow ribbon?

_____ centimeters

Personal Math Trainer

8. THINK SMARTER✚ Measure the length of each object. Which object is longer? How much longer? Explain.

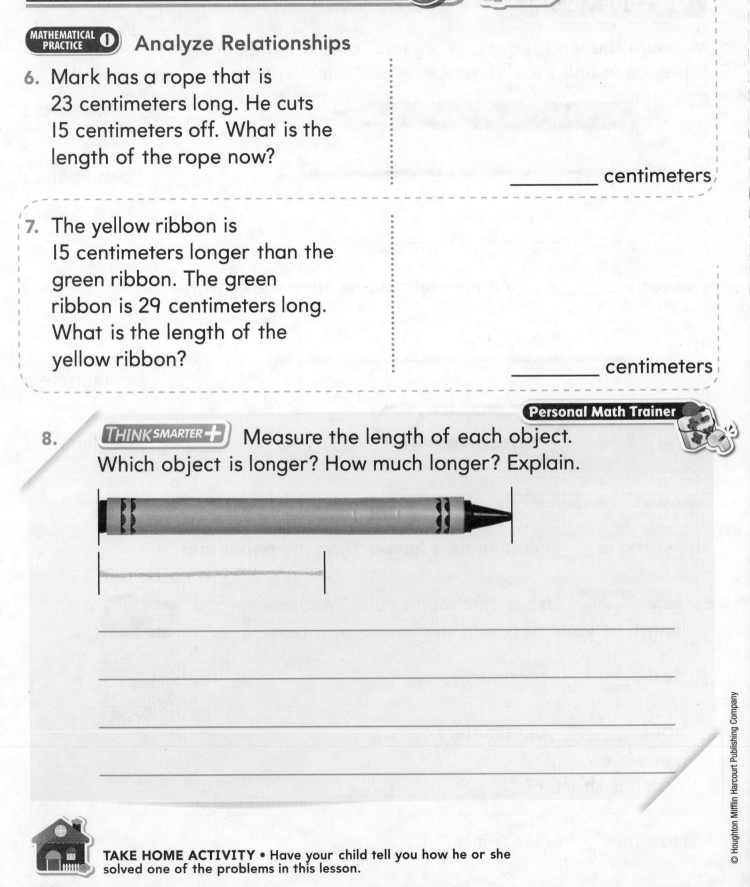

_____

_____

_____

_____

TAKE HOME ACTIVITY • Have your child tell you how he or she solved one of the problems in this lesson.

Measure and Compare Lengths

Common Core **COMMON CORE STANDARD—2.MD.A.4**
Measure and estimate lengths in standard units.

Measure the length of each object. Write a number sentence to find the difference between the lengths.

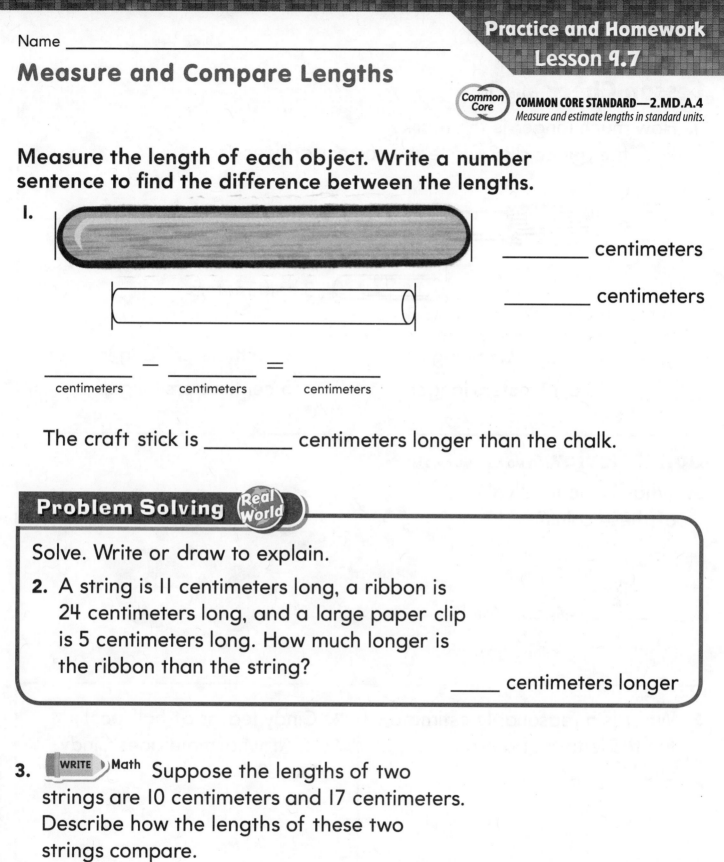

1.

_____ centimeters

_____ centimeters

_____ − _____ = _____
centimeters   centimeters   centimeters

The craft stick is _____ centimeters longer than the chalk.

Problem Solving Real World

Solve. Write or draw to explain.

2. A string is 11 centimeters long, a ribbon is 24 centimeters long, and a large paper clip is 5 centimeters long. How much longer is the ribbon than the string?

_____ centimeters longer

3. **WRITE** Math Suppose the lengths of two strings are 10 centimeters and 17 centimeters. Describe how the lengths of these two strings compare.

_____

_____

Lesson Check (2.MD.A.4)

1. How much longer is the marker than the paper clip? Circle the correct answer.

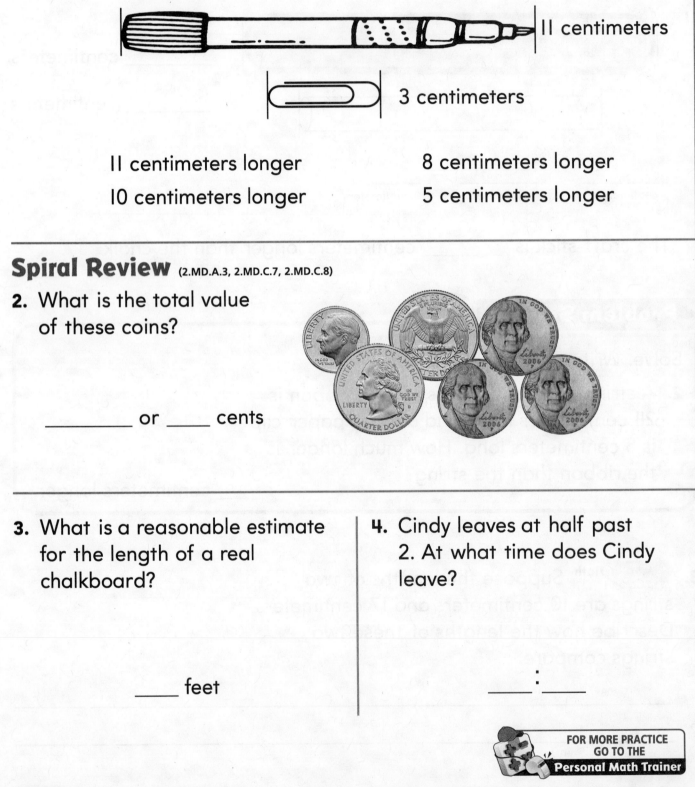

11 centimeters

3 centimeters

11 centimeters longer   8 centimeters longer

10 centimeters longer   5 centimeters longer

Spiral Review (2.MD.A.3, 2.MD.C.7, 2.MD.C.8)

2. What is the total value of these coins?

_____ or _____ cents

3. What is a reasonable estimate for the length of a real chalkboard?

_____ feet

4. Cindy leaves at half past 2. At what time does Cindy leave?

_____ : _____

FOR MORE PRACTICE
GO TO THE
Personal Math Trainer

✓ Chapter 9 Review/Test

1. Michael uses unit cubes to measure the length of the yarn. Circle the number in the box that makes the sentence true.

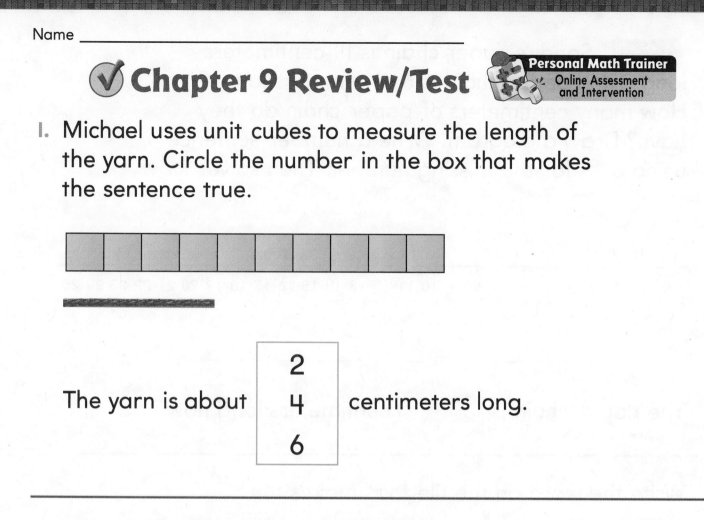

The yarn is about
2
4
6
centimeters long.

---

2. The paper clip is about 4 centimeters long. Robin says the string is about 7 centimeters long. Gale says the string is about 20 centimeters long.

Which girl has the better estimate? Explain.

_____

_____

 Assessment Options
Chapter Test

3. **GO DEEPER** Sandy's paper chain is 14 centimeters long. Tim's paper chain is 6 centimeters long. How many centimeters of paper chain do they have? Draw a diagram. Write a number sentence using a 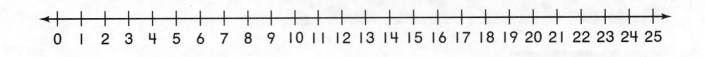 for the missing number. Then solve.

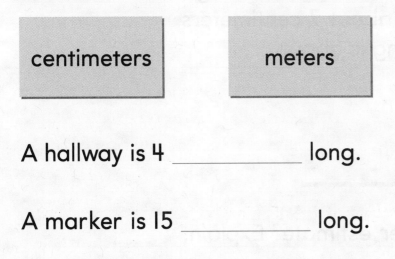

The paper chain is _____ centimeters long now.

4. Write the word on the tile that makes the sentence true.

centimeters	meters

A hallway is 4 _____ long.

A marker is 15 _____ long.

A toothpick is 5 _____ long.

A sofa is 2 _____ long.

5. Estimate the length of a real car. Fill in the bubble next to all the sentences that are true.

○ The car is more than 100 centimeters long.

○ The car is less than 1 meter long.

○ The car is less than 10 meters long.

○ The car is about 20 centimeters long.

○ The car is more than 150 meters long.

Personal Math Trainer

6. **THINK SMARTER +** Measure the length of each object. Does the sentence describe the objects? Choose Yes or No.

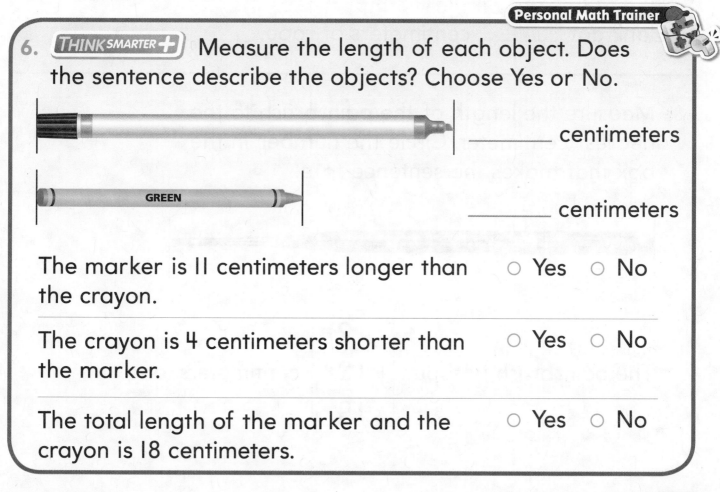

_____ centimeters

GREEN

_____ centimeters

The marker is 11 centimeters longer than the crayon.	○ Yes	○ No
The crayon is 4 centimeters shorter than the marker.	○ Yes	○ No
The total length of the marker and the crayon is 18 centimeters.	○ Yes	○ No

7. Ethan's rope is 25 centimeters long. Ethan cuts the rope and gives a piece to Hank. Ethan's rope is now 16 centimeters long. How many centimeters of rope did Hank get from Ethan?

Draw a diagram. Write a number sentence using a ■ for the unknown number. Then solve.

0 1 2 3 4 5 6 7 8 9 10 11 12 13 14 15 16 17 18 19 20 21 22 23 24 25

_____

Hank got _____ centimeters of rope.

---

8. Measure the length of the paintbrush to the nearest centimeter. Circle the number in the box that makes the sentence true.

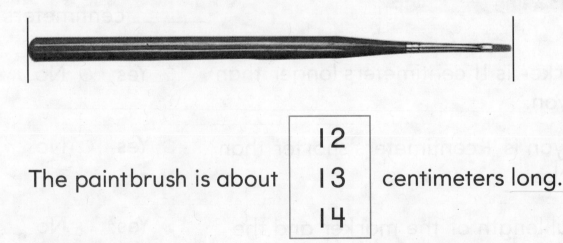

The paintbrush is about | 12 / 13 / 14 | centimeters long.